HOW PEOPLE LIVE

IN THE DESERT

Nancy Dickmann

BROWN BEAR BOOKS

Published by Brown Bear Books Ltd
4877 N. Circulo Bujia, Tucson, AZ 85718, USA
and
Studio G14, Regent Studios, 1 Thane Villas, London N7 7PH, UK

Text: Nancy Dickmann
Design Manager: Keith Davis
Children's Publisher: Anne O'Daly

Library of Congress Cataloging-in-Publication Data
Names: Dickmann, Nancy, author.
Title: In the Desert / Nancy Dickmann.
Description: Tuscon, AZ : Brown Bear Books Ltd., [2025] | Series: Fast Track: How People Live | Includes bibliographical references and index. | Audience: Ages 5-7 years | Audience: Grades K-1 | Summary: "How people live in and near deserts all around the world"– Provided by publisher.
Identifiers: LCCN 2023052211 (print) | LCCN 2023052212 (ebook) | ISBN 9781781219713 (library binding) | ISBN 9781781219775 (paperback) | ISBN 9781781219836 (ebook)
Subjects: LCSH: Deserts–Juvenile literature. | Arid regions climate–Juvenile literature. | Architecture, Domestic–Arid regions–Juvenile literature.
Classification: LCC GB612 .D53 2025 (print) | LCC GB612 (ebook) | DDC 551.41/5–dc23/eng/20231220
LC record available at https://lccn.loc.gov/2023052211
LC ebook record available at https://lccn.loc.gov/2023052212

The photographs in this book are used by permission and through the courtesy of:
Cover: Shutterstock: Afzal Khan Maheen. Interior: Dreamstime: Roberto Caucino 11; iStock: kbels 4-5bk, 12-13bk, Privizer 14; Shutterstock: ActiveLines 8-9bk, 16-17bk, AleCasa77 12, Chantelle Bosch 4, Catay 13, Brent Coulter 20t, Artur Didyk 18, Embyr Graphics LLC 23, Fotografo de los Andes 20b, HennadiiH 1inset, Elena Istomina 3inset, Anton Ivanov 9, Jemastock 1, 2-3, 20-21bk, 22-23bk, 24, Jingcq 10, Litchi Cyril Photographe 8, N MacTavish 10-11bk, Benny Marty 17, Abdullah Nalmamari 7, Elena Podolaya 19, Porcupen 20-21c, posztos 6, Eleanor Scriven 21t, SpreadtheSign 6-7bk, 14-15bk, 18-19bk, sunsinger 21b, Fredy Thuerig 15, Irina Tokareva 5, Vector Tradition 24br, Di Vincenzo 21cr, Grace Wangui 16.
t-top, b-bottom, l-left, r-right, c-center, bk-background
All other artwork and photography © Brown Bear Books.

Brown Bear Books has made every attempt to contact the copyright holders. If you have any information about omissions please contact: licensing@brownbearbooks.co.uk.

The website addresses in this book were valid at the time of going to press. However, it is possible that contents or addresses may change following publication of this book. No responsibility for any such changes can be accepted by the author or the publisher. Readers should be supervised when they access the Internet.

Words in **bold** appear in the Words to Know on page 23.

Manufactured in the United States of America
CPSIA compliance information: Batch#AG/5659

Contents

Not a Cloud in the Sky

The sun beats down. It heats the sand and rocks. This is a desert. It looks empty and **barren**. But some plants and animals survive here. People live in deserts too.

Desert animals are **adapted** to survive the dry conditions.

WOW!

The biggest desert is in Antarctica! The land is covered in ice and snow. But not much new snow falls.

A desert is a place that gets very little rain. Many deserts are in hot places. Some are in cold places. Some are sandy. Others have rocks and mountains. They're all very dry.

Weather and Climate

Deserts get little rain. Most days are sunny.
Many deserts are hot in the day.
They get very cold at night.

People build fires to stay warm at night.

Strong winds blow sand across the desert.
There are also sand and dust storms.
The sand can block the sun.

Desert Homes

Some desert houses are modern.
They have air conditioning.
Old homes also stay cool.
They have covered courtyards.

Some desert people move around. They live in tents.

WOW!

Shibam is an old city in Yemen. Its buildings are made of mud brick. Some are seven stories tall!

Many desert houses have thick walls.
They are often made of mud brick.
They stay cool, even on hot days.

Food and Drink

It is hard to grow **crops** in the desert. There is not enough rain. But people farm at **oases**. They grow dates and figs. They grow wheat and olives.

The water in an **oasis** comes from underground.

Some desert people keep camels.
They move around to find grass.
The animals give them milk and meat.

Clothes

In hot deserts, you need to stay covered.
The sun can burn your skin.
Desert clothes are light.
They are loose-fitting.

In cold deserts, people wear thick layers to stay warm.

WOW!

The Tuareg are sometimes called the "blue men of the Sahara." Their clothes were dyed blue. It rubbed off on their skin!

People often wrap a cloth around their head.
It shades them from the sun.
The long cloth can cover the face.
It protects from blowing sand.

Jobs

Many deserts have gas and oil. These **resources** are below the ground. People work on **oil wells**. Deserts are also rich in **minerals**. Some people have jobs digging them out.

Deserts can be windy. People work on wind farms. The windmills turn wind into power.

There are towns and cities in deserts. People work in stores. There are teachers and doctors. Other people help **tourists**. They show them around. They work in restaurants and hotels.

Getting Around

Some desert people ride on camels.
Camels are adapted to life in the desert.
Their wide feet don't sink into the sand.
They can go for days without water.

Camels can carry heavy loads. A line of camels is called a caravan.

Many deserts are very empty.
Service stations are far apart.
Drivers need to be prepared.
They must carry extra fuel and water.

Games and Sports

Desert sand can pile up into big **dunes**. People slide down the dunes on boards. It's like snowboarding, but on sand!

People ride on ATVs. The tough tires are good on sand.

This is the Festival of the Sahara.

It happens each year. It celebrates desert life.

People race on camels and horses.

They enjoy music and dancing.

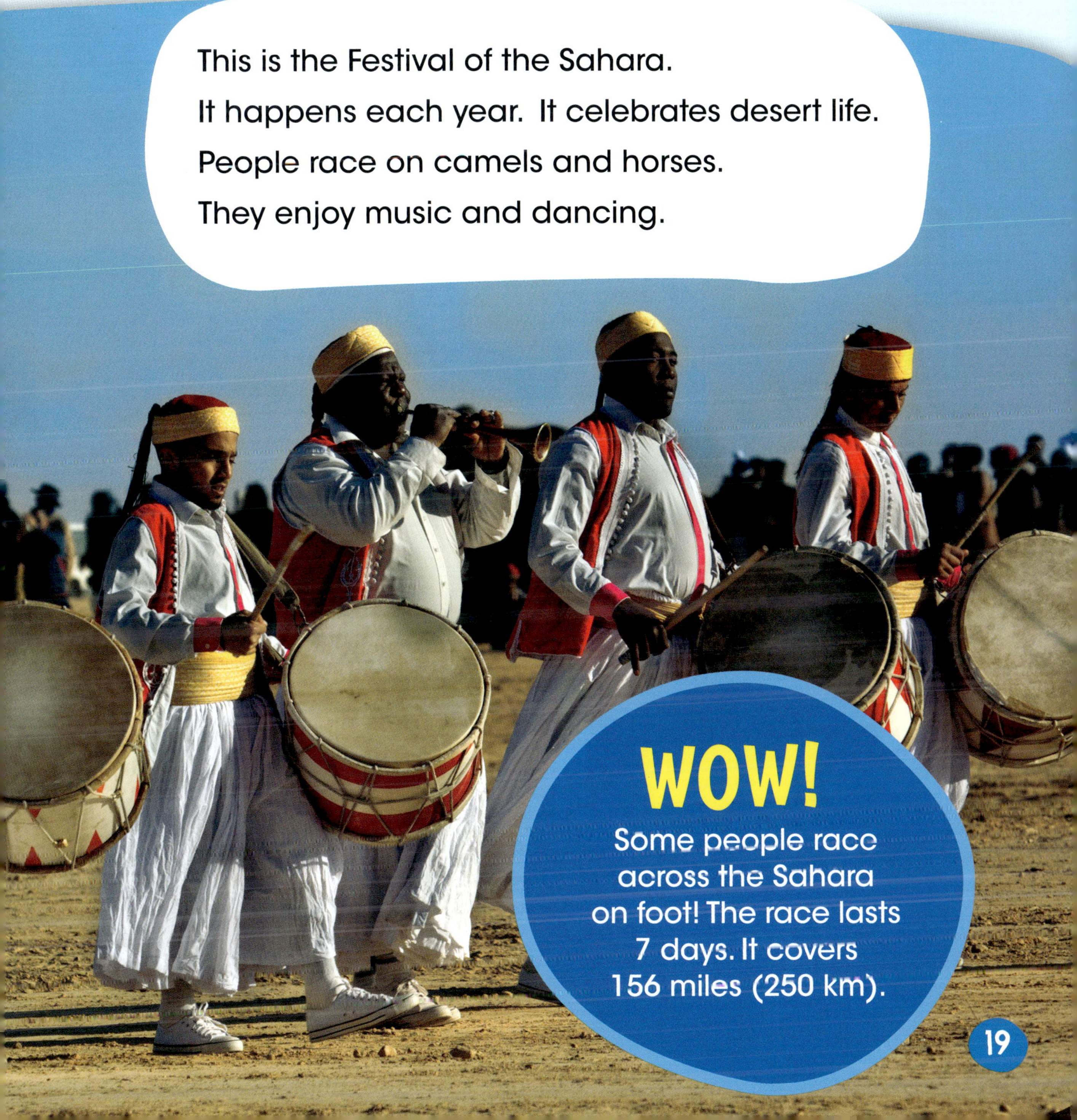

WOW!

Some people race across the Sahara on foot! The race lasts 7 days. It covers 156 miles (250 km).

Where in the World?

There are deserts all around the world. Here are some of them.

Desert covers large areas of the southwestern USA.

Chile's Atacama Desert is a great place for stargazing.

The Gobi is a cold desert that covers parts of Mongolia and China.

Deserts in Australia are often called "the outback." Not many people live there.

The Sahara is the world's largest hot desert. It stretches across Africa.

Activities

Do you live in a desert? If not, where is the nearest desert? Find it on a map. How long would it take you to get there?

About one billion people live in deserts. Some live in modern cities. Others have a more traditional life. Research a group of desert people. How do they survive?

Many different animals survive in the desert. Choose one and learn more about it. How has it adapted to the dry conditions?

Find Out More

Websites

dkfindout.com/us/earth/deserts/what-is-an-oasis/

kids.britannica.com/kids/article/desert/346108

kids.nationalgeographic.com/nature/habitats/article/desert

Books

All About Deserts Christina Mia Gardeski, Capstone Press 2018

Desert Biome Elizabeth Andrews, Cody Koala 2022

Living in the Desert Joanne Mattern, Rosen Publishing, 2021

Words to Know

adapted having a change in body or behavior that helps a plant or animal survive in its habitat

barren having little or no plant life

crops plants that people grow to eat

dune a hill of sand piled up by the wind

minerals non-living substances found in the ground, which can be useful for making things

oasis (plural **oases**) an area of fertile ground in the middle of a desert where plants can grow

oil well a hole dug to reach oil underground

resources materials that occur naturally which people can use or sell

tourist a person who travels to a place on vacation to see the sights

Index